(including Stanton Harcourt)

by Peter Davies

RAF Abingdon was planned in the mid-twenties as a base for two single-engined day bomber squadrons, to have four aircraft sheds and permanent technical and domestic accommodation. The site lay two miles (3.2 km.) west of Abingdon and six miles (10 km.) south of Oxford, and the planned location of an airfield so close to Oxford was not accepted without comment. On 21 November 1929 the Under Secretary of State for Air was asked *"if his attention had been called to the protests against the establishment of a great aerodrome for military purposes in the immediate vicinity of Oxford, and if so whether His Majesty's Government propose to continue a project which will go far in destroying the amenities of the University and in increasing the danger to which it might be subjected in time of war?"*.

In reply, the Under Secretary stated that he was aware of the protests and that *"...the site at Abingdon was, however, only decided upon after an exhaustive reconnaissance had failed to discover any feasible alternative in the vicinity, and the aerodrome now under construction is an essential part of the scheme for the defence of this country"*.

Abingdon Council received a letter on 22 May 1929 from Sir Alan Cobham, asking whether there was a large field nearby for use in connection with a flying tour which was being organized. A reply was sent, notifying the company of the purchase by the Air Ministry of a large piece of land near the village of Shippon. Until the opening of the airfield as RAF Abingdon, it was always referred to locally as Shippon Aerodrome.

After a few alterations to the original plans, work commenced in 1929. On 21 May 1930, the Council notified the Air Ministry of damage being done to local roads by vehicles carting materials for construction of the airfield. Four Type A Aeroplane Sheds, which were the RAF's first permanent end-opening hangars and which became the standard hangar for the period, were constructed to the north of the domestic and technical site. In the annexe to one was the air pilotage office and on the airfield side a watch office of unknown pattern.

Sir Alan Cobham's visit did not take place until 29 April 1932, when as part of his National Aviation Day campaign the display took place alongside the Fringford road, almost opposite the new airfield. Sir Alan mentioned that Abingdon was one of the first places that he had visited with an air show back in 1919, when he was with Berkshire Aviation Company.

It was not until 1 September 1932 that RAF Abingdon opened, in the charge of Wg. Cdr. G. W. Robarts MC, five other officers, one Warrant Officer, eight NCOs and 42 airmen, as part of Wessex Bombing Area. The first unit to arrive was 40 Sqn., which had left its base at Upper Heyford on 31 July for the Armament Training Camp at Catfoss, from where its Gordon biplane bombers flew to Abingdon on 8 October.

The first buildings were officially taken over on 10 October, and at the end of the month Station Flight was established with an Atlas, an Avro 504N and a DH Moth. On 2 November Oxford University Air Squadron (OUAS) arrived in formation, an Atlas leading Avro 504Ns. OUAS had been formed at Upper Heyford in October 1925 and this move to Abingdon was to begin an association with the Station which would last, with breaks, for more than sixty years. The average combined strength of SHQ, 40 Sqn. and the Station Flight for the first year was 24 officers, four WOs, 56 NCOs, 185 airmen and nine civilians.

Abingdon became the headquarters of the Central Area organisation on 16 November 1933, the unit having been formed at Andover six weeks earlier. Central Area's function was to control the growing number of bomber airfields in central England in those days before the 'role Commands' were set up.

RAF Abingdon opened its gates to the public for the first time on 24 May 1934, when the Empire Air Day attracted 1622 adults and 485 children. The entrance fees were respectively 1/- and 6d (5p and 2.5p), the proceeds, then as now, going to the RAF Benevolent Fund. Flying took place between 14.00 and 18.15 and a static display included a fully-equipped Gordon, 504Ns and a range of guns, cameras and parachutes.

On 1 June 1934 Abingdon gained its third resident unit when 15 Sqn. was reformed as a day bomber squadron, to be equipped with Harts.

Hawker Hind K5465 of 40 Sqn., based at Abingdon. This aircraft joined the squadron on 10 March 1936 and was transferred to 185 Sqn. on 7 June 1938. [I. Morgan]

Rather than conform with the norm, the CO, Wg. Cdr. T. W. Elmhurst, decided to have the squadron number painted in Roman numerals and on 18 June flew a Hart to Hendon for the forthcoming display, sporting letters XV. By February 1935 15 Sqn. was training on the Otmoor bombing range, 17 miles (27 km.) to the north east. In April the squadron was tasked with training crews of other units in instrument flying, for which three additional Harts were taken on charge.

In 1935 Empire Air Day took place on 25 May on similar lines to that of the previous year, and attracted 3200 spectators. Having shown its prowess in bombing skills and airmanship, 15 Sqn. was now considered the foremost light bomber squadron in the RAF and on 29 June,

nine of its aircraft led the Light Bomber Wing at the Hendon air display. The squadron again took the lead in a flypast on 6 July, when as part of King George V's Jubilee Review of the Royal Air Force they flew over Duxford, witnessed by HM the King, Queen Mary, the Prince of Wales and the Duke and Duchess of York.

At the end of 1935, 40 Sqn. started to replace its Gordons with Harts. During this change-over period, on 9 November, Gordon K2720 and Hart K4371 of 40 Sqn. suffered a mid-air collision and crashed at Abingdon. On 7th January 1936, 104 Sqn. was reformed from 'C' Flight of 40 Sqn., to which it remained attached until July. Ten days later 98 Sqn. was similarly reformed from 'B' Flight of 15 Sqn. In March, 40 Sqn. and its progeny 104 re-equipped

A formation of 40 Sqn. Hinds -K5423 closest to the camera aircraft with K5470 leading - flying from Abingdon in 1936. [I. Morgan]

Hind K5463 of 15 Sqn. was delivered new on 22 April 1936 but on 4 September that year was hit by a gust of wind while landing and overturned, to be Struck Off Charge on 16 October. [I. Morgan]

with Hinds. Two months later its 'B' Flight formed the nucleus of 62 Sqn., which departed two months later to Cranfield, followed by 104 Sqn. to Hucknall in August.

In February 1936, three of 15 Sqn's. 'C' Flight Harts were disposed of in advance of receiving Hinds, six of which were delivered during March and six in April. 'C' Flight's Hinds then formed the basis of 98 Sqn. when it was reformed at Abingdon on 17 February, although it did not become independent until May. 15 Sqn. took part in the winter air exercise during the month with 18, 40, 57 and 101 Sqns. Due to bad weather, raids were only carried out on the 18th, 15 Sqn. making the only successful strikes. Although outwardly similar to the Hart, the Hind was an interim replacement to enable squadrons to be formed before the introduction of the new Battle and Blenheim monoplanes, which would take some time to reach squadron service.

On 20 March 1936 2 (Bomber) Group was formed at Abingdon as part of Central Area and on 1 May was joined by 1 (Bomber) Group, both Groups becoming part of the new Bomber Command on 14 July 1936.

The 1936 Empire Air Display took place on 23 May, 15 Sqn. 'A' Flight with Hinds displaying close formation flying and 'B' Flight demonstrating dive bombing with Harts. This display was a forerunner of military displays as we know them today, as the static line-up had aircraft from other Stations — a Bulldog, a Heyford, a Moth and an Overstrand and some civil aircraft belonging to Phillips & Powis of Reading, in addition to a stripped-down Hind, a Hind 'equipped for war' and a Tutor.

Locally-built Pou-de-Ciel (Flying Flea) G-AEEX was flown at Abingdon in 1936 by a resident officer. In May, the sole Clarke Cheetah Monoplane, G-AAJK, became based at

Gloster Gauntlet Mk.II K5280 spent time with a number of units after delivery to 10 FTS on 28 April 1936, and is seen here as a visitor to Abingdon. The numeral 2 on the fuselage may indicate participation in the Hendon display. [I. Morgan]

K1748 was a Fairey Gordon delivered to 40 Sqn. on 3 September 1931 and used by the squadron until November 1935. [I. Morgan]

Abingdon with its owner. It was a year for civil aircraft, as Percival Vega Gull VP-KCC left Abingdon at 18.50 on 4 September to fly an east-west solo transatlantic attempt, piloted by well known South African aviatrix Mrs Beryl Markham, who was successful in her attempt.

Late in January 1937, 2 (Bomber) Group HQ moved to Andover and 15 Sqn. spawned yet another squadron when five of its officers and 26 airmen formed the nucleus of 52 Sqn. Seven Hinds were collected from Sealand during the month and 52 Sqn. moved to Upwood two months later.

The 1937 Empire Air Day on 29 May attracted 6000 visitors and visiting aircraft included an Anson, a Hendon, a Moth, an Overstrand, a Swordfish and a civilian Miles Hawk. On 18 June Oxford UAS went to Ford for ten days for the annual training camp, Cambridge UAS replacing it.

During Special Bombing Trials in June, 15 Sqn. dropped 2500 dummy bombs on Abingdon airfield and the Otmoor range. The squadron was to re-equip with the Battle monoplane bomber, and the first aircraft, K9214, was collected on 13 June 1938. This single aircraft was used to convert the pilots during the remainder of June and July. June also saw the reformation of 106 Sqn. from 15 Sqn's. 'A' Flight. During July, 15 Sqn. disposed of its final three Hinds and 40 and 106 Sqns. also re-equipped with Battles. On 29 July 62 Sqn. took part in the mass flypast at the Hendon Air Day.

35 Sqn. from Lympne arrived with Hinds

Hind K5440 of 15 Sqn. ran into soft ground at Abingdon while taxying on 4 February 1937 but was repaired to fly again. [I. Morgan]

The Gloster Gladiators of 65 Sqn. from Hornchurch arrived at Abingdon on 2 February 1938 for an affiliation exercise. This rare picture shows the FZ code worn by the squadron, which began to re-equip with Spitfires in March 1939, a month before such two-letter codes were promulgated! [John Chapman]

on 18 September for a fourteen-day attachment for bombing trials. The Fleet Air Arm made its first appearance at Abingdon on 3 November when the Swordfish of 825 Sqn. arrived, followed next day by 802 Sqn. with Nimrods and Ospreys. Both squadrons were from HMS Glorious, and remained until 17 January 1938, when they returned to the carrier.

A Flight of 65 Sqn. Gladiator fighters from Hornchurch joined 40 Sqn. on 2 February 1938 for an affiliation exercise. In March, 185 Sqn. was reformed from 'B' Flight of 40 Sqn., using Hinds until June, when it also equipped with Battles. Another new Sqn., 106, was formed from a Flight of 15 Sqn. on 1 June, temporarily equipped with six Hinds until it received Battles, of which it had eight by 11 July.

Empire Air Day in 1938 was on 28 May and the static display included not only an Anson, Gauntlet, Hart, Harrow, Hector, Magister and Overstrand but new types in the shape of a Battle and a Blenheim. A new treat for the local civilians this year was that certain buildings were open for inspection. In June OUAS went to Ford once more for its summer camp and CUAS again took its place. On 1 September, 106 and 185 Sqns. departed to Thornaby with their Battles, making room for similarly-equipped 103 Sqn. to move in from Usworth next day.

With the threat of war in Europe looming, 'War Crisis' was declared in September 1938, and 15 and 40 Sqns. were instructed to be operationally fit by the 15th. On 24 September all ranks were recalled from leave and a number of officers moved in, mainly from the Staff College at Andover, to take up posts with HQ 1 Group, which caused overcrowding. Following the signing of the Munich Agreement by Neville Chamberlain on 28 September, the tension eased. However, to see what its aircraft were capable of, 40 Sqn. sent six Battles on an endurance trial with a full bomb load on 25 October. Panic over, life on the Station returned to normal but on 9/10 November a tactical exercise took place involving all squadrons, and 122 hours were flown over the period. On the last day of the year the Station strength was 73 officers and 625 airmen.

During the late 1930s civilian air traffic control was in its infancy and military control was almost non-existent. At the suggestion of AOC-in-C Bomber Command, six Military Area Control Centres were set up in January 1938 to control en-route traffic. Each was to be manned by two Flying Control Officers on airfields selected for their strategic importance. Abingdon was one and the others were Boscombe Down, Leuchars, Linton-on-Ouse, Mildenhall and Waddington.

Although not an official title, in the spring of 1939 15 Sqn. became known as "Oxford's 'own' squadron". On April Fools' Day, 103 Sqn. moved to nearby Benson, while 15 and 40 Sqns. each became affiliated with an Air Defence Corps unit. Although no Empire Air Day event was held at Abingdon that year, on 8 May 15 and

One of the many Whitleys used by 10 OTU at Abingdon and Stanton Harcourt was K9025, which joined the unit on 10 May 1940 but was transferred to 19 OTU on 25 August that year.

40 Sqns. took part in mass formation flights elsewhere. 40 Sqn. was affiliated with the Borough of Abingdon on 19 July 1939, while 15 Sqn., which now had twenty-four Battles, was practising on the Porton Down bombing range.

In July 1939 the squadrons practised operational raids and cross-country flights, including some over France, which was visited by nine 40 Sqn. aircraft on the 11th. Following a Home Defence Exercise in August, most Station personnel went on leave, but on 24 August mobilisation commenced, and they were recalled. 15 and 40 Sqns. became part of the Advanced Air Striking Force, formed at Abingdon on 24 August 1939 to control 71, 72, 74, 75 and 76 Wings. On 1 September a number of civilian transport aircraft flew in and by nightfall were camouflaged. At 10.00 next morning they began ferrying ground personnel to

Bethenville in France. An hour later the Advance Air Party of 15 Sqn. departed in four flights of four aircraft at ten-minute intervals, and that of 40 Sqn. left between 12.00 and 15.45, one aircraft force-landing 15 miles (24 km.) north of Dieppe. Next day, Kidlington, referred to as Thrupp, was requisitioned as a satellite airfield. War was declared with Germany the following morning.

63 Sqn., a training unit at Upwood, moved to Abingdon with Battles and a handful of Ansons on 7 September. Its stay was short, as after only ten days it moved on to nearby Benson. On 9 and 10 December some 15 and 40 Sqn. Battles made a brief re-appearance at their former base when they landed from France on the way to Wyton, where they were to re-equip with Blenheims.

Training was to be Abingdon's role for the

HRH the Duchess of Gloucester leaving the 6 Group HQ building after her visit in July 1940.
The architecture of this fine building can be clearly seen.
[Mrs. Joan Sylvester]

ABINGDON including STANTON HARCOURT

His Majesty King George VI walking on the airfield at Abingdon during his visit in August 1940. He was accompanied by the Station Commander, Gp. Capt. Massey DSO MC. Behind the party can be seen the tail of a Whitley of 10 OTU, carrying full-height fin markings. [Mrs. Joan Sylvester]

next six years, due to its location away from the operational zone in the east of England. On 17 September 1939, two squadrons, 97 and 166, moved in from Leconfield with twelve Whitley bombers and four Ansons each to form 4 Group Pool at Abingdon. Next day, 1 Group Pool also came into being, comprising 52 and 63 Sqns., each with 18 Battles and six Ansons.

As with all training units, accidents were the inevitable price to pay when intensive use was made of aircraft often well past their prime. It was not long before accidents began to occur to the Abingdon-based units. Whitley K7185 of 166 Sqn. overshot on landing from a night navigation sortie on 11 October and K7203 suffered a tyre burst and undercarriage collapse after undershooting on a night landing on 6 November. It was the turn of 97 Sqn. when on 8 November Whitley K7225 undershot on landing at Abingdon, hitting a tree and a house 300 yards (275 m.) from the runway with the loss of three lives; then on 18 December K7260 crashed upside down at Kempton after hitting trees in bad visibility, again with the loss of three lives.

Another 97 Sqn. Whitley, K7255, hit trees at nearby Boars Hill on 11 February with three on board and crashed, killing both pilots and being destroyed in the ensuing fire. 166 Sqn. Whitley K8960 stalled off a turn after take-off on 12 March, hitting the west wing of the Officers'

Mess and crashing on the domestic site, with the loss of all three on board. The upper floor of the mess was gutted and a civilian mess steward was rushed to hospital with severe burns from which he died five days later.

In accordance with new policies, the two squadrons comprising 4 Group Pool merged on 8 April 1940 to form 10 Operational Training Unit (OTU), a unit which was to operate at Abingdon for the next six years. 97 Sqn. became 'A' and 'B' Flights equipped with Whitleys and Ansons, and 166 Sqn. became 'C' and 'D' Flights, equipped with Whitleys. On its formation the OTU had an establishment of 54 Whitleys and 18 Ansons, but that very day the pilot of Whitley K8957 lost control in cloud and crashed near Stratford-on-Avon. K9007 struck a machine gun post in a force landing on the airfield after the undercarriage jammed on 5 May and K7227 belly landed on approach to Pershore after losing an engine four days later. On 23 May, a visiting Honington-based Wellington of 9 Sqn., L7777, crashed when it overshot on landing.

A suitable area for a satellite airfield was now selected seven miles (11 km.) north-west of Abingdon, across the Thames at Stanton Harcourt in Oxfordshire. Geo. Wimpey & Co. Ltd. was chosen as contractor and began work in May 1940 on the £210,000 project. Stanton Harcourt is located on a gravel terrace between

WAAFs from Abingdon did their bit to stimulate recruitment into the service. These girls were photographed on arrival at Reading in June 1940. [Mrs. Joan Sylvester]

the Rivers Thames and the Windrush and is noted as a site of historic interest due to the Devils Quoits, a Neolithic stone circle. Once a religious site of some importance, there are many burial sites in the area. Stanton means stone town and is a reference to the stones, of which there were originally 35 in a 245-foot (75 metre) circle, but all but three were destroyed by medieval farmers and it was not until 1940, during construction of the airfield, that the remaining three were buried, some 4500 years after their erection.

During the afternoon of 15 June, 73 aircraft of the Advanced Air Striking Force made re-fuelling stops at Abingdon, where a temporary HQ had been set up for them following Operation 'Dynamo', the withdrawal from France. Next day sixteen more staged through and a few more stragglers arrived on the 17th.

In 1940 and 1941 7 Anti-aircraft Co-operation Unit (AACU), based at Castle Bromwich, operated a detachment at Abingdon, using a mixed bag of aircraft, one of which, Dragon Rapide G-ADIL, crashed during a night landing at Abingdon on 1 July. A most unusual former civilian aircraft impressed into RAF use was Messerschmitt Bf.108B-1 Taifun AW167, which arrived at Abingdon on 14 July for use by the Station Flight. Formerly D-IJHW, this four-seater had been based at Croydon for use by the German Embassy until the outbreak of war, when departure for its homeland was thwarted by a contrived puncture.

On 19 July 1940, 10 OTU was honoured by a visit and inspection by HM King George VI. Two days later, 10 OTU had its first taste of operational flying, when three Whitley Vs made a 'nickelling' sortie (dropping propaganda leaflets) over Abbeville, Amiens, Rouen and Le Havre, each aircraft carrying 1500 lb. (680 kg.) of leaflets and two 250 lb. (114 kg.) bombs, the latter carried with the express orders that they were only to be dropped on enemy airfields. Many such sorties were flown in the coming months.

Five Wellingtons of 3 Group landed at Abingdon when returning from operations on 4 August. On 5 August 1940, a 10 OTU Whitley was shot up and forced to land at Squires Gate near Blackpool. The Luftwaffe was not to blame, however, as the attacker was a Hurricane from Silloth! Using Abingdon as an operations base on the night of 12/13 August, twenty Whitleys of 4 Group attacked targets in the Turin area, and two nights later four more went for the same target, one aircraft failing to return. Three more attacks were made by 4 Group from Abingdon before the end of the month.

On 16 August, during construction of one of the runways, Stanton Harcourt received a visit from three Luftwaffe aircraft which strafed the site, leaving eleven casualties, among them five Irish workers who were killed and four more who died later from wounds. Next day anti-aircraft defences were put in position. Damage to the airfield was only minor and on 3 September 10 OTU began night flying there. Three runways of an unusual layout, forming a triangle, were tarmac-surfaced, lengths being 06/24 4800 ft. (1463 m.); 12/30 and 00/18 both 3300 ft. (1006 m.). There were 23 heavy bomber hardstandings and two hangars, one a B.1 and the other a T.2. The original control tower and operations room combined was built to 15898/40 pattern, the standard for bomber satellite airfields, but was replaced eventually by a standard building to pattern 343/43.

Whitleys of 10 OTU 'C' Flight moved into Stanton Harcourt on 10 September to concentrate on night flying training. Sgt. Bob Pratt was posted to 10 OTU in September 1940 after qualifying as a W/Op, and during the first few weeks flew in Ansons and Whitleys, including bombing at the nearby Marcham range. Flying was from both Abingdon and Stanton Harcourt and in October Bob underwent air gunner training before being posted onto operational Whitleys of 78 Sqn. at Dishforth. September also saw the re-opening of OUAS for short ground training courses to the standard of the RAF Initial Training School 'Wings' syllabus, and during the war years handled about 2000 students.

On the last day of September 10 Sqn. Whitley P4957 just made it to Abingdon after losing an engine over the Channel when returning from an operation. Stanton Harcourt received another visit from the Luftwaffe on 3 October, when three bombs were dropped, damaging a runway. On the 11th Whitley N1526

Stanton Harcourt, seen through a camera carried by a No.1 PRU aircraft on 14 Febrauary 1942.

lost wing fabric and crashed out of control at Akeman Street near Witney, and burned with the loss of six crew.

In November the control staff at Abingdon moved into a new control tower of 5845/39 pattern with a meteorological section. With major extensions, this 'Villa'-type building continued in use until replaced in 1960.

It was Abingdon's turn to be targeted for an attack when on 3 December at 16.50 six bombs were dropped from 2000 ft. (610 m.). Two fell on the Station and four close by, causing minor damage to buildings and injuring four people, one seriously. On 9 December three Wattisham-based Blenheim IVs of 107 Sqn. operated out of Abingdon to make attacks on Vannes, Rennes and Brest. Three days after Christmas Topcliffe-based Whitley P5111 of 77 Sqn. ran out of fuel and crashed on Abingdon's instructional centre when attempting to land. During December experiments were carried out to devise a method of taxiway lighting using railway lamps borrowed from the Great Western Railway!

An increase in bad-weather flying by all commands in the RAF, and in particular that of Bomber Command, led to the formation of Blind Approach Training Flights (later retitled Beam Approach Training Flights) to give pilots practice in the use of blind landing systems. Initially, priority for installation of these systems was given to Bomber Command airfields, due to increasing losses of aircraft trying to find their bases after returning from raids on Germany in inclement weather. Between 1935 and 1939 the RAE had evaluated six different systems of American, British, Dutch and German origin at Abingdon. The German-designed Lorenz system proved the most effective and was selected for use by the RAF, British rights having been obtained by Standard Telephones & Cables in 1936. 1 BAT Flight was formed at Abingdon on 12 January 1941, affiliated to 10 OTU. BAT equipment was installed on the airfield but it was not until 2 February that the Flight's first aircraft of five, Whitley K8990, was collected. Over the next ten days a working party from West Drayton fitted them with blind approach receiver equipment and on 24 January four pilots commenced the first week-long course.

1941 opened with bad weather and a sad incident on 17 January when Whitley N1494 crashed on the Wooton Road after an engine cut due to icing while on approach in a snow storm. Two crew had safely baled out but four were killed in the ensuing crash. Doris Hickton was on the staff of the Station intelligence library at Abingdon, which was housed in a wooden hut near the control tower. The Station Intelligence Officer, as chairman of the Committee of Adjustment, was responsible for ensuring that the belongings of deceased personnel were sorted, the kit returned to a central depository and personal items to the family. This onerous task was often passed to Doris, who recalls sitting at her desk surrounded with little bags containing the effects of young airmen that she had known, tears streaming down her face. Doris was later sent to Stanton Harcourt to set up a Station intelligence library, and as the only WAAF there at the time she was billeted at the manor house in the village.

Although Stanton Harcourt had been in use for a few months, building work was still far from complete, work on the domestic site not starting until 27 January 1941 and the technical site until 11 February. When completed, there were six dispersed sites in the village, including all accommodation and messing facilities, and at a later date the bomb dump was situated in open country south-east of the airfield. A shortage of Whitleys forced 'C' Flight at Stanton Harcourt to disband in February 1941, and it was replaced by 'A' Flight, which converted crews to the Whitley.

March saw more attacks on Abingdon; on the 12th, two bombs dropped on the bomb dump failed to explode and were successfully defused, the only casualties being two 1 BAT Flt. Whitleys damaged by bomb splinters. Two days later, Whitley N1429 was returning from a night training exercise over Marcham bombing range, and was followed by a Ju.88 which dropped its load on the airfield, fortunately without much damage. On 15 March, when the Station was still recovering from the previous day's excitement, HM the King of Norway visited 10 OTU with the C-in-C of Bomber Command, Air Marshal Sir Richard Pierse KCB DSO AFC. Only a week later, a stick of 29 bombs was dropped on Abingdon, 28 falling outside the airfield, leaving one fatality and some HQ 6 Group buildings damaged.

Stanton Harcourt was still having its share of accidents in 1941; Whitley N1411 crashed a mile south-west of the airfield after an overshoot on 21 April and later in the year N1429 hit trees on approach to land on 10 October and K8981 crashed on landing on 5 November. On 13 June, Abingdon Station Flight Tiger Moth N6673 failed to recover from a loop and crashed at nearby Clifton Hampden.

RAF Abingdon continued to fare well with Royal and VIP visitors when on 23 May HM the King and HM the Queen with Princesses

This Whitley Mk.V of 10 OTU, Z9466 [ZG:Q], overshot its landing at Stanton Harcourt on 22 June 1944. [via R. C. Sturtivant]

Elizabeth and Margaret were guests of 10 OTU. During May, the instructional centre which had been devastated by the crash the previous December was rebuilt to include two new Bombing Teachers. Two Lysander IIIs, a new type to the OTU, were received on 17 August and modified to tow target drogues.

During September 1941 there seems to have been confusion over the parentage of Mount Farm airfield, which had recently been relinquished by 1 PRU at Benson, supposedly to 15 OTU at Harwell, but at the same time it was referred to as 10 OTU Mount Farm Detachment. To confuse matters further 10 OTU's Operations Record Book stated "Personnel on the satellite (Mount Farm) are to be regarded as posted to 10 OTU, but temporarily attached to 15 OTU Harwell for duty at Mount Farm." This situation prevailed until 25 January 1942, when Mount Farm was officially handed over to Harwell.

Both Abingdon and Stanton Harcourt received many visiting aircraft, mostly training types, but as both airfields were able to support bombers, the numbers of these returning from operations in need of fuel or urgent attention was increasing. During October 1941 Stanton Harcourt was the departure point for a detachment of thirteen 104 Sqn. Wellingtons bound for Malta, from where they, and more of the type which passed through later, were to attack targets in Libya, Sicily and Italy before

moving on to Egypt. Four Whitleys arrived on 26 December to take a consignment of bullion worth £500,000 to the Middle East, three departing on 28 December and one next day.

In October, Stanton Harcourt and Mount Farm were tested for suitability of installing BAT equipment, and 1 BAT Flight was renumbered 1501, the precursor of almost sixty such units which would be formed during the war. HQ Bomber Command notified 1501 BAT Flight on 10 May 1942 that it was to re-equip with Oxfords, the first of which was delivered with some haste two days later, and by the 15th another four had arrived.

In addition to the 'nickelling' sorties which had been carried out since July 1940, by the end of 1941 'gardening' (mine laying) sorties were also flown by the OTU. On 20 May 1942, fourteen Wellingtons and two supporting Ansons of 20 OTU arrived at Stanton Harcourt from Lossiemouth for operations over Cologne, the ground crew being ferried in by a Harrow.

During 1942, 10 OTU provided aircraft and crews for the thousand-bomber raids, the first on the night of 30/31 May, when twenty-one Whitley Vs bombed Cologne and returned safely. Between June and August the OTU participated in a further nine raids, three to Bremen, two to Dusseldorf, two to Essen and two to Hamburg. For its part, 10 OTU supplied 148 crews, 103 of which reached their target and 39 aborted due to bad weather or mechanical problems. Six OTU aircraft failed to return from these raids. After a year of operational flying, Sgt. Bob Pratt had returned to 10 OTU 'A' Flight at Stanton Harcourt the previous November as a W/T instructor. On 26 May in N1374, captained by Flt. Lt. Daniels, they positioned at Abingdon, from where an air test was carried out each day until they departed at 23.10 on 30 July to bomb Cologne. After being airborne for five hours fifty minutes they landed at Hampstead Norris without brake pressure, returning to base in an Anson. The same aircraft and crew bombed Essen on 1 June. Sgt. Pratt was the WOp/AG on Whitley N1391 when it lost an engine due to a glycol leak while on a cross country on 2 September and he had to bale out near Nuneaton in Warwickshire, leaving the aircraft to force-land near Bedworth.

In the summer of 1942 10 OTU was tasked with supplying aircraft and crews to take part in the Battle of the Atlantic. Their role would be to undertake twelve-hour anti-submarine patrols over the Bay of Biscay, for which task additional fuel tanks were fitted. On 4 August the first twenty-six Whitleys left

Abingdon for St Eval in Cornwall. The first encounter with an enemy aircraft was on 28 August, when a Whitley piloted by Plt. Off. Chisolm was attacked by a Ju. 88. After the ensuing exchange of fire the Ju .88 was seen to dive into the sea. Next day Sgt. McClelland made the first attack on a submarine, fighting off a Ju.88 in the process. It was not until 19 July 1943 that the last patrol was made by an OTU aircraft. During this period, in 1848 sorties 16,864 operational hours had been flown, 91 U-boats were sighted and 54 attacks made. The price had been high, though, as no less that thirty-three 10 OTU aircraft had been lost to enemy fighters or mechanical problems when over the sea. A Book of Remembrance in the church at St. Eval records over 800 names of those who lost their lives when serving there, 91 of them 10 OTU crews.

Reorganisation of 10 OTU in November left Abingdon with 'C', 'D' and 'G' Flights and Stanton Harcourt with 'A' and 'B' Flights. Len Warner, who spent his first Christmas in the RAF at Stanton Harcourt, remembers his billet in the middle of the village and the cookhouse and 'ablutions' on the way to the airfield. The

The Station Intelligence section at Abingdon in 1942. Several model aircraft used for recognition purposes hang from the ceiling, while maps and charts cover the walls. [Doris Hickton]

cookhouse doubled as the NAAFI, where dances were held on Saturday nights after the washing-up had been completed! It is on record that the village 'came to life' with the arrival of the RAF, with dances on the camp and in the village hall. What must have been of great interest to both aircrew and groundcrew was a visit to Abingdon on 8 December by 1426 (Enemy Aircraft Circus) Flight with an He.111, a Ju. 88 and a Bf.110. These aircraft were demonstrated in the air and were available for close scrutiny on the ground.

On 12 January there was a buzz of excitement at Stanton Harcourt when Liberator AL504 'Commando' flew in from Lyneham, piloted by Gp. Capt. van der Kloot, for operation 'Static'. AL504 had a special VIP fit, and early next morning it departed for the Casablanca Conference, carrying Sir Winston Churchill to meet President Roosevelt and Stalin.

1682 Bomber (Defence) Training Flight was formed at Stanton Harcourt in June, equipped with a few Tomahawk I and IIa single-seat tactical reconnaissance and ground attack fighters for fighter affiliation training with 10 OTU's Whitleys. Tomahawk I AH822 suffered an engine fire and belly-landed in a field near Oxford on 17 August 1943 and AH860 crashed on overshoot at Abingdon on 24 February 1944, two days before the Flight moved to Enstone to be affiliated with 21 OTU, based at Moreton-in-Marsh..

During April 1943 1501 BAT Flt. had moved to Stanton Harcourt, No. 108 Course commencing the same day, but at the end of December the Flight was disbanded, having served its useful purpose. Four Whitleys made 'nickelling' sorties on 7 January 1944, two over Paris and two over Versailles. Flying times for January were 919.40 day and 987.20 night for the Whitleys, 125.25 day and 99.05 night for the Ansons, 69.06 day and 22 night for the Martinet and 119.25 and 10.45 day, respectively for the Tomahawks and Oxfords.

On 4 February 1944 a 300 (Polish) Sqn. Wellington crashed at Stanton Harcourt on return from a minelaying sortie. A Halifax from each of three squadrons, 433, 466 and 640, landed at Stanton Harcourt after an attack on Stuttgart on 21 February and three days later the airfield received three Halifaxes and a Lancaster returning from an attack on Schweinfurt. Martinet MS781 suffered engine failure just after take-off from Stanton Harcourt on 18 March, bounced on the runway and ran into a stream. Another Martinet, JN284, lost brake pressure and ran into iron railings on 5 April and on 17 May the same aircraft stalled after take-off from Abingdon and dived into the ground, killing the pilot.

By July, some Hurricanes were on the strength of the OTU, as Hurricane MW346 stalled during a camera gun affiliation with a Whitley and dived into the ground at Mount Farm on 11 July. The long serving Whitley was now near the end of its useful life and during July the OTU began to re-equip with Wellington Xs, the first arriving on 29 June.

On 5 August, Grove-based USAAF C-47 118599 flew into Stanton Harcourt from France with wounded servicemen from the Continent. The first OTU Wellington incident occurred on the night of 23/24 August when the starboard undercarriage of LN543 collapsed on landing. Next night, possibly Stanton Harcourt's busiest, 10 OTU despatched one Wellington and ten Whitleys on Bullseye sorties, and there were twelve weather diversions following raids, eleven Halifaxes of 424, 425 and 427 Sqns. and a 620 Sqn. Stirling returning from a 'special mission'. During the night of 26 August, 424 Sqn. Halifax LW113 made a forced landing with a full bomb load after aborting a mission to Brest due to problems with the port inner engine. That month, 34 Wellingtons were fitted with dual controls, bringing the unit up to three-quarters OTU strength. 'A' Flight had retained Whitleys to complete training on the type but 'B' and 'C' were now fully equipped with Wellingtons.

Although at Stanton Harcourt there were concrete runways, at Abingdon there was only a grass landing area which became both waterlogged and slippery after rain, with a 50 ft. (15 m.) wide concrete perimeter track. It was not until April 1944 that preparations were put in hand for construction of two tarmac runways, and on 20 March all flying was transferred to Stanton Harcourt for the duration of the work. On 26 August, 4200 ft. (1280 m.) of runway 01/19 was completed and ready for use and at 12.25 an OTU Hurricane became the first aircraft to land on it, but it was not until October that the full 6000 ft. (1830 m.) length was completed. There was only one subsidiary runway, 09/27, 4800 ft. (1463 m.) long. Construction of the runways almost doubled the airfield area and resulted in the realignment of some roads and the demolition of some houses. There were now thirty hardstandings, six spectacle-type and twenty-four frying-pan.

October 1944 saw the departure of the final Whitleys and Ansons from 10 OTU. Since the formation of the unit in April 1940 370+ individual Whitleys had been on charge. Masters were now also on strength. It was not until 16

Cheerful-looking ground crews of 10 OTU pose in front of one of their charges, a Wellington Mk.X, at Stanton Harcourt in 1944. [Betty Hutton]

November that daylight flying was resumed at Abingdon and by early 1945 most of the unit had returned there.

On 1 December 1944, RAF Abingdon's personnel establishment was 1120 RAF and 274 WAAF and Stanton Harcourt's 709 and 101 respectively, but there was an increase next day, when a Polish Flight from 18 OTU at Finningley arrived, bringing the unit's strength up to that of a full OTU. The night of 14/15 January 1945 saw four OTU Wellingtons involved in 'Sweepstake' sorties over the North Sea. On the night of 8/9 April, sixteen Lancasters of 207 and 619 Sqns. landed at Abingdon after attacking Lutzkendorf and 16 Lancasters of 227 Sqn. landed at Stanton Harcourt after taking part in the same raid.

Victory in Europe was celebrated on 8 May with dances at Abingdon and Stanton Harcourt, a dinner in the Officers' Mess at Stanton Harcourt and a party in Abingdon's Officers' Mess. On 14 May, Lancasters of 227 Sqn. again visited Abingdon, when two aircraft

flying under Operation 'Exodus' landed, each with twenty-four repatriated prisoners of war who continued by road to Oakley. During June and July 'Cook's Tours' gave an opportunity for operational and non-operational personnel to fly over the Continent to see the effects of the bombing offensive.

With the war almost at an end, 11 July was a black day for the OTU when three Wellingtons were lost. NC715 lost power on take-off and raised its undercarriage to stop, but two aircraft on night navigation exercises were not so fortunate. Both flew into the ground after flying out of clouds, LP873 at Kineton in Warwickshire and NC714 at Much Wenlock, Shropshire. Losses of 10 OTU crews over the past five years, during training or when on the St. Eval detachments, or on thousand-bomber raids, had been considerable, but were probably similar to other OTUs. Only a month after VJ-Day, Abingdon held its first post-war 'RAF at Home', when more than 10,000 people were able to see

The bomb-aimer training building at Stanton Harcourt in remarkably good condition when photographed in 1983 after nearly forty years of disuse. [author]

at close quarters the types that had become household names to them.

By the summer of 1945 Stanton Harcourt had ceased to be used by 10 OTU, although Martinet JN655 suffered an engine failure between there and Abingdon on 25 October and crashed two miles (3.2 km.) from Stanton Harcourt, without injury to the pilot. All flying ceased at Stanton Harcourt on 23 November and, now redundant, it closed officially two days later. Abingdon's strength at this time was 229 officers, 294 SNCOs and 577 other ranks (RAF) and four officers, sixteen SNCOs and 331 other ranks (WAAF).

Two Spitfire XVIs were added to the the OTU's inventory in January 1946 and as the year progressed more were taken on charge. On 12 January the AOC, AVM J. A. Gray, CB CBE DFC GM presented the 10 OTU badge to the CO, Gp. Capt. A. King-Lewis. The badge consisted of a practice bomb and the torch of learning, crossed to represent the Roman numeral X, with the inscription *'Ex Scientia Vires'* (Strength through knowledge). However, having survived longer than most OTUs, 10 OTU's task now at an end, and all flying ceased on 10 September 1945, leaving Abingdon strangely quiet.

Abingdon's role changed once more when the Station was handed over to Transport Command on 24 October 1945. The first unit to arrive under the new command was 525 Sqn., which moved its Dakotas in from Membury at the end of the month. Its task of operating the mail and newspaper service between the UK and British bases on the continent continued for only a month before it was renumbered 238 Sqn. Dakotas were retained, eight being on strength in March 1947, and by May some Horsa gliders and at least one Oxford were on charge. December 1946 saw the arrival of a second Dakota unit when 46 Sqn. moved in from Manston. During the year 4 Group HQ and its Communications Flight were also set up.

An affray in the airmen's mess at 18.00 on New Year's Day 1947 between Jamaican and British airmen resulted in two of them sustaining injuries. As a result of a Court of Inquiry, all Jamaican personnel were posted away! That winter was a long and bitter one and at a completely snowbound farm at Gagingwell near Enstone the farmer requested assistance from the RSPCA. Contact was made with the RAF and on 7 March a drop of food for the livestock was made by 238 Sqn. Dakota KJ829. Two years later KJ829 joined Rolls Royce at Hucknall, where it was fitted with Dart turboprops, becoming the Dart development aircraft for the next thirteen

Seen loading food for livestock to be dropped at Gagingwell Farm, near Enstone, on 7 March 1947 is Dakota Mk.IV KJ829 [WF:V] of Abingdon-based 238 Sqn. Many villages were isolated and roads cut by heavy snow that spring, prompting such air-drops. This Dakota went on to see service with Rolls-Royce as the Dart engine test-bed. [via R. Harrington]

York MW183 served with Station Flight at Abingdon for a time before moving to Bassingbourn for similar duty. Eventually it was sold, becoming G-AMUU of Air Charter, who used it on trooping flights with new serial XD668.

years. 46 Sqn. crews made several special ferry flights during the summer of 1947, two in July to Athens to deliver Dakotas for the Royal Hellenic Air Force. In September, 48 and 238 Sqns. were joined by two other Dakota squadrons, 53 from Netheravon and 77 from Broadwell, for training for Operation 'Longstop', a mass parachute operation, the first such operation to take place since the war. The combined squadrons' task involved keeping 48 aircraft constantly airborne, the actual operation taking place on 22/23rd September. On 20 September, spectators at the Battle of Britain display were treated to a mass landing of 48 of these aircraft in ten seconds! In September, 46 and 238 Sqns. began a detachment at Vienna, operating until 7 November. 27 Sqn. was reformed as another Dakota unit (on paper) in early November, and a nucleus of aircrew was transferred from 46 Sqn.

The Dakota's stay at Abingdon was not to last, as in November 46 Group HQ informed the Station that its Flying Wing and personnel would move to Oakington and would be replaced by 40, 51, 59 and 242 Sqns., equipped with Yorks. On 24 November, 27, 30 and 46 Sqns. took up residence at Oakington, and two days later fifteen Dakotas, five Horsas, one Tiger Moth and one Oxford departed for their new base. The final Dakota schedule left Abingdon on the last day of November and next day the Station was transferred from 46 Group to 47 Group. On this day, 40 and 59 Sqns. reformed at Abingdon, 242 Sqn. moved in from Oakington, and a week later the final York squadron, 51, arrived from Waterbeach. The first York incident occured on 16 December, when MW301 lost both starboard engines on approach and belly landed on the outskirts of the airfield. There were no injuries and the cause was put down to a malfunctioning valve and the intervention of 'sod's law' when the airfield lighting failed at the same time.

In February 1948, the desired flying training was set at 100 hours per month per squadron, and route flying was to be split

Abingdon-based York MW204 [KY:F] of 242 Sqn. with its crew, some of them taking no notice of the photographer!

RAF Abingdon, as seen on 12 April 1946 from a F20" camera aboard a 541 Sqn aircraft.

between the squadrons on a weekly basis; the month's total flying time was 2,647 hours. The Yorks were operated as long-range transports for passengers and freight, making regular flights to the Middle East, Far East and India.

The division of Berlin into sectors by the occupying powers in 1945 was soon to lead to the first post-war crisis in Europe. A growing rift between the Allies and the Soviet Union resulted in the city being cut off from the rest of Europe by road and rail, the only means of entry to deliver the necessary 4500 tons of supplies each day being by three air corridors. On 28 June 1948, the USAF commenced transport ops into the city, starting what was to become known as the Berlin Airlift. Soon the RAF became involved, establishing a Transport Wing at Wunsdorf in west Germany to control Dakotas and Yorks. Abingdon's squadrons were put on

standby on 29 June and on 1 July 47 Group notified them that all route and training flights would cease from that day. Five Yorks and eight spare crews departed for Wunsdorf at once and by 6 July 21 of Abingdon's Yorks with 35 crews were at Wunsdorf to take part in what became known as Operation 'Plainfare'. On 27 July, 59 Sqn. York MW311 swung on take-off and suffered an undercarriage collapse at Abingdon, writing the aircraft off. The month ended with 313.35 hours route flying and 2130.50 hours being logged on Operation 'Plainfare'.

Although the Station was stretched with the supply of Berlin, the 1948 Battle of Britain display took place on 18 September, when members of the public were treated to an interesting variety of aircraft on the ground and in the air. Formations displayed included Lincolns, Spitfires, Tempests, P-80 Shooting

Unusually smart Valetta C.2 VX576 of 30 Sqn. sports both the squadron badge below the cockpit and the number in a diamond on the fin.

Stars of the USAF and Seafires of 1832 Sqn. which had flown the four miles (6.5 km.) from RNAS Culham, the other side of Abingdon. A King's Flight Viking, a York, a Hastings, the Theseus engine testbed Lincoln and parachute drops from a Halifax were also seen.

In April 1949 OUAS took its nine Tiger Moths to Kidlington, providing Abingdon with much-needed space for its York operations. With Brize Norton's impending change in role to Flying Training Command, the Transport Command Development Unit (TCDU), which carried out trials with the transportation and delivery of airborne loads and paratroops, received notification that it was to move to Abingdon between 27 June and 2 July. An advance party flew into their new base on 27 June, followed by the main party two days later.

By the end of June, 90% of personnel, with kit and bedding, and over 90,000 lb of equipment had made the journey by air, using one Halifax, one Dakota and one Valetta!

On 1 March 1950, the Air Transport Development Flight formed within Flying Wing at Abingdon, but more significant was the formation on 10 June of 1 Parachute School on the closure of 1 Parachute & Glider Training School at Upper Heyford. That unit had used a variety of aircraft, but at Abingdon the task was concentrated on ten Dakotas C.4, six Horsa gliders and an Oxford. Its role was to provide basic static-line parachute training from both balloons and aircraft for the Regular and Territorial armies.

The work of ferry units at this time was still a very important service in support of RAF

Hastings C.2 WD476 of 24 Sqn., based at Abingdon between May 1953 and January 1957.

WD500 was the first of only four Hastings C.4 aircraft and was used by 24 Sqn. to carry VIPs world-wide during the early 1950s.

units around the globe, and for the delivery of British aircraft to Commonwealth and other overseas air arms. In March 1951 the Overseas Ferry Unit moved in from Chivenor to continue this task. A typical month's work — July 1952 — for what had become 1 (Overseas) Ferry Unit was as follows: deliveries to MEAF were three Mosquito 36s (of which one crashed), twenty-two Vampire 9s, two Meteor T.7s and an Anson; to FEAF were seven Hornet F.3s (one crashed), eighteen Meteor F.8s and two Mosquito 34s; to 2nd TAF were nine Vampire FB.5s, three Vampire 9s, two Tiger Moths and an Auster. In addition six Mosquitos were delivered to Jugoslavia, three Vampire 5s were returned from MEAF and a Dakota and a Valetta were returned from FEAF. Deliveries of older types are exampled by a Lancaster to MEAF in August, a Tempest to 2nd. TAF in September, a Proctor from Schipol in October and one Wellington and four Spitfires to Malta in November. Early in November 1952, 1 OFU was divided into 1 and 2 Flights, which on 1 January 1953 became 1 and 3 (Long Range) Ferry Units, the conversion training part becoming the Ferry Training Unit.

During January 1953, Transport Command Air Support Flight lost Hastings TG602 near Shallufa in Egypt when it lost an elevator and dived into the ground, with the loss of nine lives. Abingdon-based Hastings made regular detachments to Shallufa in the mid-fifties, in support of the large forces still retained in the Canal Zone.

1 and 3 (LR) Ferry Units were upgraded to squadron status on 1 February, becoming 147 and 167 Sqns. Next day the official hand-over ceremony of the first Sabres for the RAF took place at Abingdon. Eight aircraft delivered in December were presented by the people of Canada to the people of Britain by the High Commissioner for Canada, Mr. N. A. Robertson and accepted for the RAF by Rt. Hon. the Lord de L'Isle and Dudley VC, Secretary of State for

Air. Three earlier examples were also delivered to Abingdon for the Ferry Training Unit.

In April 1953 the Ferry Training Unit and 167 Sqn. moved to Benson and 147 Sqn. followed within a few weeks. More Hastings arrived in May, when 24 and 47 Sqns. moved in from Topcliffe. Sadly, when visiting Abingdon on 22 June, Lyneham-based 53 Sqn. Hastings WJ335 crashed on take-off, killing seven crew members. A minor alteration to the parachute school was made on 1 November 1953, when it became 1 Parachute Training School.

Five instructors from 1 PTS took part in the World Parachuting Championships that year, and during the year 'free-falling' came into the syllabus. 47 Sqn. inaugurated a Flying Club in December with Hawk Trainer G-ADWT. By April 1955 the Club had fifty members and in May the Hawk Trainer was replaced by a Tiger Moth became G-AOBH. Over the next five years the Tiger was used for flying instruction, glider towing at Weston-on-the-Green and Bicester and free-fall parachuting.

On 28 April the Freedom of Entry into the Borough of Abingdon was conferred upon the Station. At a ceremony held in the Corn Exchange, the Mayor, Dr. G. Fitzgerald O'Connor, presented the casket containing the Title Deed to the Station Commander, Gp. Capt. S. P. Haggar, who signed the Freedom Roll.

In 1952, the Ministry of Supply had placed an initial order for the Beverley C.1 for RAF Transport Command, an aircraft designed as a versatile short-range transport for the carriage of large and/or bulky loads, able to operate from short strips. The seventh production aircraft and the largest aircraft yet to enter RAF service, XB265, was the first of the type to join a squadron when it was delivered to 47 Sqn. on 12 March 1956. 47 Sqn. had just phased out the Hastings, and by June had received its complement of eight Beverleys. The sheer size of the aircraft caused headaches where maintenance

Just about to touch down at Abingdon is Beverley XB286 [Z] of 47 Sqn., which carried RAF Air Support Command titling.

was concerned. Ground equipment for engine servicing was inadequate and none of Abingdon's hangars was large enough to accommodate the aircraft, so all servicing had to be carried out in the open. A new hangar for the Beverleys was only at the design stage.

The introduction of the Beverley was not without teething problems, but in May the first route flight was made to Malta, a week later regular flights to Cyprus began and by the end of the month twice-weekly flights to Wildenrath were inaugurated. Training with the Army commenced in June; one such trial during the month involved landing on the grass airfield at nearby Watchfield and disgorging 98 troops in under two minutes.

The Beverley made its operational debut in November 1956, when the Suez Crisis made good use of its bulk freighting capability. One aircraft flew twenty-one sorties from Nicosia, Cyprus to Port Said's Gamil airfield, delivering over 270 tonnes of fuel in drums for the Army. Perhaps the most notable flight during this conflict was the delivery of a 47 ft (14.3 m.) long radar scanner from Abingdon to Nicosia. After the cease-fire, the aircraft assisted with the return

A type not seen in the air for many years is the Belvedere twin-engined helicopter. Here XG455 [B] of 72 Sqn., based at Odiham, is seen while visiting Abingdon for the Battle of Britain open day in 1963.

of troops and supplies to the UK. Beverleys were also used in the supply of medical supplies and dried milk to Vienna during the Hungarian crisis. During the first year of Beverley operations 47 Sqn. carried 2763 passengers and over 10,500,000 lb of freight in and out of Abingdon.

53 Sqn., which had flown Hastings at Lyneham, moved into Abingdon on New Year's day 1957 to re-equip as the second Beverley squadron and five days later 24 Sqn. departed for Colerne. The first Beverley to be lost was XH117 of 53 Sqn, which crashed at Drayton, near Abingdon, on 5 March.

In July, the United Kingdom Mobile Air Movements Squadron (UK MAMS) was formed on the Station as part of the Transport Command Air Movements Development Unit. UK MAMS personnel were trained on all aspects of loading and unloading all British aircraft and helicopters and also some of other nationalities.   31 April 1959 was a day that Abingdon's Beverley groundcrew would remember — the day that the new 'F' hangar was handed over for use, allowing the leviathans to be worked on in the dry. Another new building, handed over for use on 24 July 1960, was a new Type 2548c/55 control tower on the north-west side of the airfield.

The threatened annexing of Kuwait by Iraq in June 1961 saw 53 Sqn. operating in the Middle East in support of MEAF aircraft. During the year the 1 PTS Free Fall Parachute Display Team were named 'The Falcons', a name they proudly bear today at Brize Norton. At about this time the RAF Sport Parachute Club was formed and in 1963 acquired two ex-Fleet Air Arm Dominies from Lossiemouth for £400.

At the end of June 1963 53 Sqn. disbanded on merging with 47 Sqn., which then became the only UK-based Beverley squadron. 53 Sqn. had flown over 6500 sorties from Abingdon and an estimated 2000 when on detachments, logging an estimated 30,000 Beverley flying hours.

A tragic accident stunned the Station on 6 July. Colerne-based Hastings TG577 had just taken-off with a crew of five and 38 paratroops bound for the jump site at Weston-on-the-Green when the crew reported 'sloppy controls' and requested a priority landing. The Hastings did not, however, reach Abingdon but crashed with the loss of all lives near Dorchester, ten miles (16 km.) south-east of the airfield. The accident was attributed to failed elevator bolts.

January 1966 saw the formation of the Belfast Servicing Flight, housed in the 'F' hangar, although this was not large enough to take the fin and rudder and was certainly too low for a jacked-up aircraft. In the summer of 1967 the Flight moved to Brize Norton, where it had the luxury of the huge 'Base Hangar'. UK MAMS was granted Squadron status on 1 May, becoming an independent unit within Air Support Command.

Dominie, NF847, was sold on 5 February 1963 to become G-ASIA, owned by F. B. Sowerby & Partners and one of a pair of DH89's (the other being G-ASFC) intended to be operated by the RAF Sport Parachute Club. It had seen service with the FAA from 1945 and had been used by assorted RNAS Station Flights. Here two Cub Scouts are inspecting it at Abingdon's 1963 open day.

Andover C.1 XS603 of 46 Sqn. touching down at Abingdon on 15 June 1968.

A military freighter development of the very successful Avro 748 twin turboprop airliner was ordered by the RAF in the early sixties as the Andover C.1. It initially equipped the newly-formed Andover Training Unit and 46 and 52 Sqns., which reformed at Abingdon in December 1966. 46 Sqn. was to remain at Abingdon but after a few weeks working-up period 52 Sqn. went to Seletar, Singapore.

Transport Command was renamed Air Support Command in January 1967. During that year cracks were found in wing centre section fuselage joints on some Beverleys, and with the arrival of the new Hercules imminent all squadrons operating the type were disbanded, 47 Sqn. on 31 October. Two aircraft, XB269 and XB290, soldiered on for another month with ATDU, making a final flypast over HQ Air Support Command at Upavon on 6 December on the way to 27 MU at Shawbury for scrapping.

Over the years, Abingdon's Engineering Wing has been responsible for restoration and maintenance of many historic aircraft, including those of the Battle of Britain Memorial Flight. This expertise began in the mid to late 1960s, when Avro 504K E449, Spitfire 1a AR213/G-AIST and the Afghan Hind passed through the Wing's capable hands.

Immaculate Chipmunk WZ877 of 6 Air Experience Flight is prepared to take another ATC cadet airborne from Abingdon.

AIRSHOW '68!!

Now in the hands of the Shuttleworth Trust at Old Warden, the Hind (Afghan) was shown in Afghan Air Force markings at Abingdon in June 1968. [J. F. Hamlin]

With Abingdon's control tower in the background, Sea Balliol T.21 WL732 took part in the 50th Anniversary of the RAF in June 1968 while still in service with the Fleet Air Arm. [J. F. Hamlin]

1968, the 50th Anniversary of the Royal Air Force, saw many celebrations in commemoration, culminating in the Royal Review at Abingdon on 14 June. HM the Queen, the Duke of Edinburgh, the Queen Mother, the Duchess of Gloucester and Princess Marina inspected more than sixty aircraft types spanning the fifty years, and a huge hangar exhibition depicting the history of the RAF. In the afternoon some 160 aircraft old and new took part in a flying display opened by 31 Jet Provosts flying in a Royal Cypher 'E II R' formation. Next came vintage aircraft led by the Vickers Gunbus replica followed by Spitfire, Hurricane, Mosquito and Lancaster, to modern displays and formations of current types. Making a spectacular sight and sound was a formation of V-bombers, six Victors leading eighteen Vulcans. Abingdon's own 'Falcons' parachute display team performed a perfect drop, a pre-view of the RAF of the future was demonstrated by a Nimrod and a Harrier and a rousing finale was given by nine scarlet Gnats of the 'Red Arrows'. Next day the display was repeated in front of 70,000 people when the Station was open to the public, a day to be remembered by all those lucky enough to have been there.

ATDU had been at Abingdon under various guises since 1949, but in 1968 it formed part of the new Joint Air Transport Establishment (JATE), remaining at Abingdon with another section at Old Sarum. April 1970 saw the return of the Belfast Servicing Flight from Brize Norton, 'F' hangar having been modified in advance, with dormers to accommodate the fins of the Belfasts. In September, the Station bid farewell to both 46 Sqn. and the Andover Training Squadron when they departed for Thorney Island. Shortly afterwards, the Air Support Command Examining Unit (ASCEU), which later became No. 46 Group Examining Unit, took up residence. In January 1971, JATE was re-organised with Abingdon as its parent Station.

A very significant civil aviation event took place that June when 67 aircraft and crews from nine countries gathered for the London to Victoria (British Colombia, Canada) Air Race, organised to commemorate the centenary of the Canadian state. The race was started by the Canadian High Commissioner at 06.32 on 1 July. The winner was a German Merlin, second was an Irish Twin Comanche and third an American Cessna 310.

Seen at Abingdon on 29 June 1971 was Merlin D-IBMG,(race entrant 42) the winner of the London to Victoria air race. [W. J. Bushell]

The merger of Air Support Command into Strike Command in 1972 placed Abingdon under the new Command. August 1973 saw the welcome arrival of two Chipmunk-equipped units, London University Air Squadron (LUAS) and 6 Air Experience Flight (AEF), both moving across from White Waltham. In October, LUAS became the first unit in the RAF to equip with the Bulldog. 46 Group Examining Unit moved to Upavon in October and December saw the departure of long-time resident unit UK MAMS to Lyneham.

Early in 1974 rumours of massive defence cuts were rife and Abingdon's future looked bleak. Published on 19 March, the White Paper stated that Abingdon would remain open,

although its role would be changed. It would come under Support Command and the Belfast Servicing Flight, 1 PTS and JATE would move to Brize Norton. Two Maintenance Units (MUs) were to take their place, 60 MU from Dishforth and 71 MU from Bicester. 60 MU had formerly been a Servicing & Storage Depot and 71 MU was the only remaining Repair & Salvage Unit in the RAF, specialising in the recovery and salvage of crashed or damaged aircraft. It brought from Bicester a number of display and demonstration airframes, including Comet 1 (portable 'Nimrod' exhibition), two Spitfires, two Gnats, nose sections of Canberra, Vulcan, Hunter, Lightning, Buccaneer and Jet Provost and two plastic replica Jaguars.

With Abingdon's control tower in the background, Bulldog T.1 XX548 [06] of University of London Air Squadron sits awaiting its next task.

Several ex-British Airways VC-10 aircraft have been placed in open storage at Abingdon pending conversion as tankers for RAF use. Here ZD230, formerly G-ASGA, is seen, looking somewhat bedraggled. In an attempt to avoid deterioration due to the elements, VC-10s were wrapped in plastic sheeting.

With the run-down of Bicester, in February 1975 OUAS returned to Abingdon, exchanging its Chipmunks for Bulldogs. In 1975, the Belfast Servicing Flight again returned to Brize Norton, followed by JATE in December and 1 PTS in January 1976, as forecast.

Both MUs merged with Abingdon's Engineering Wing in 1976, and over the years were formed into Squadrons: Aircraft Maintenance Squadron (AMS), an Engineering Support Squadron responsible for Station Workshops and the Mechanical Transport Flight and a Repair & Salvage Squadron which carried on 71 MU's former role. The initial task was major servicing and modification programmes on Jaguars. The Repair & Salvage Squadron spent most of its time away from base, working throughout the UK and sometimes overseas. Its main responsibility was for the fixed wing aircraft of the three services, but it was also involved in the recovery of many crashed civilian aircraft, transporting them from wherever they crashed to the RAE at Farnborough for accident investigation. Another responsibility was the Battle Damage Repair School, where students from all RAF Stations were taught the techniques of battle damage repair. To enable 'hands-on' experience to be carried out, many retired airframes were used, including Canberras, Buccaneers and Harriers.

British Airways had withdrawn the Super VC-10 from service in March 1981 and most were stored for possible resale. No interest was shown until the MoD favoured the type for conversion to in-flight refuelling tankers, whereupon about a dozen were ferried to Abingdon during April and May for storage pending conversion at Filton. In the summer, Engineering Wing had its first experience with Hawks, when a line was set up for a modification

Abingdon's 1988 Battle of Britain open day was cancelled after a F-4 Phantom had crashed on the runway during the practice session. This S-3A Viking of the US Navy, 160147 [710-AJ], over-ran the runway and is seen here after recovery. [W. J. Bushell]

programme and in April 1982 Hunter overhauls for the RAF and RN were began. By late 1982, the AMS was divided into two squadrons, 1 Sqn. continuing the Jaguar and Hunter maintenance and 2 Sqn., formed for Hawk major servicing. As long-term storage seemed certain for the VC-10s, they were cocooned by the end of the year. On 1 May 1985 Field Aircraft Services Ltd. took over the maintenance of the Bulldogs and Chipmunks of the two UASs and 6 AEF. The thirtieth and last Hunter to be overhauled by AMS returned to its base at RNAS Yeovilton on 17 June 1986. During March and April 1987 four of the VC-10s were broken up. Buccaneer maintenance was added to AMS during September, the first aircraft arriving in the middle of the month. The role of the Station was now officially termed an Aircraft Maintenance and Storage Depot (AMSD), one of only two in the UK, the other, at St Athan, specialising in third line maintenance.

In 1988 the Battle of Britain 'At Home' air show was cancelled when Leuchars-based 228 OCU Phantom XV428 crashed on the runway on 23 September, the day before the show, when it failed to come out of a loop while practicing its display routine in strong winds, killing both crew. Four hours earlier, a US Navy S-3 Viking from USS *Theodore Roosevelt* had overshot through a perimeter fence, ending up on a public road. On 14 September 1989, 14 Sqn. Tornado ZD710 crashed shortly after taking-off for Bruggen in Germany. Both crew ejected safely and the aircraft came down in a field. By this time, some of the ill-fated Nimrod AEW.3 conversions had arrived for storage pending disposal, and long-term gate guardian Spitfire F.22 PK624 had moved to St Athan for storage.

Eastern bloc aircraft were now beginning to appear in museums and with collectors in the West and the RAF Benevolent Fund auction at Bentley Priory on 13 September offered two examples. The Governments of Czechoslovakia and Hungary each donated a Mig 21 Fishbed, and the Czech example, 1304, was flown direct to Abingdon from Kbely on 10 September for the Battle of Britain Display.

The latter part of 1990 and early 1991 saw the departure of all but one of the VC-10s to Filton for conversion to K.4 tankers. Gulf War veteran Tornado GR1 ZA466 arrived by low loader on 23 November for attention after sustaining heavy damage during the conflict at Tabak.

Abingdon was the recipient of a most unusual visitor on 30 January 1991, when Aeroflot cargo IL-76TD CCCP-76759 arrived and off-loaded a Sukhoi SU-26MX aerobatic aircraft, CCCP-5201, for the Proteus Display Team. It later left by road for White Waltham, where it subsequently became G-ORBY.

In March, a letter to the MP for Oxford West and Abingdon, Mr John Patten, from Lord Arran, the Parliamentary Under-Secretary of State for Defence and the Armed Forces, stated that the future of RAF Abingdon was under review. This was as a result of the Government's Options for Change policy announced in the Commons the previous July, which called for a reduction in manpower from 93,000 to 75,000. Only three months later, on 13 June, the Commons Defence Select Committee announced that RAF Abingdon might close during the planned run-down of aircraft maintenance facilities, but the writing was on the wall and 'might close down' was soon translated to 'will close down'.

As part of the run-down, Hawk servicing came to an end on 9 September, when the 161st aircraft to pass through the Hawk Major Maintenance Flight returned to 4 FTS at Valley and the Flight transferred to St Athan. There was now a rapid run-down of the Station's maintenance facilities. The final Buccaneer to pass through the hands of AMS departed on 22 January 1992, followed a little over a month later by the last Jaguar, which left on 24 February and the Czech Mig 21, which went to join its new American owner in April. The RAF Exhibition Flight with most of its aircraft, and the Aircraft Salvage & Transportation Flight left for St Athan during June, and a miscellany of complete and part airframes were either transferred to other units, sold off or scrapped. Two Nimrod AEW.3s left Abingdon by road but the remaining five were scrapped between April and November and the final Super VC-10 left by road for Filton in June.

After sixty years as an active flying Station, the last based aircraft, the Bulldogs of the University Air Squadrons and the Chipmunks of 6 AEF, parted company with Abingdon's runway for the last time on 30 July 1992 on their way to Benson. At 14.00 that day thirteen Bulldogs made a stream take-off, Wg. Cdr. Cullum leading LUAS and Sqn. Ldr. Smithson OUAS. After forming up into three 'vics' and a 'diamond four' the Bulldogs made one pass before heading south-west. The OUAS aircraft had been given the honour of being the last to depart as that unit had originally moved into the new Abingdon grass airfield in November 1932.

The days of the RAF at Abingdon were now almost at and end, but fortunately the buildings would not be left to the elements — the Army was to use them as a base for the new Royal Logistics Corps. At the end of July the airfield was deactivated and over the following months the Army made preparations for their takeover. Abingdon's long association with the RAF officially ended at 17.00 on 31 July, when the last full-time Station Commander, Gp. Cpt. Henderson, gave up his post to Sqn. Ldr. Lawrance, the unit test pilot, who filled a caretaker role until the Army took over.

On 15 December 1992 at a handover ceremony, the Army officially took over its new base. Amid tight security joint service chiefs and civic dignitaries arrived, the few remaining RAF police being supplemented by MoD police to help check visitors and cars. To the tune of the Last Post played by Jnr. Tech. Carter, the RAF ensign was lowered for the last time during a

Major servicing was carried out in the huge postwar hangar at Abingdon, Jaguars being one of the types dealt with, as seen in this 1987 picture. [RAF Abingdon]

Vigilant T.1 motor-glider ZH191 [UD] is one of several used by 612 Gliding School at Abingdon in 1998 to provide tuition for Air Cadets. [author]

brief but moving ceremony. Sqn. Ldr. Lawrance then handed his charge over to his Army successor, Col. Cross. A mixed Army/RAF guard of honour marched past the dais while Sqn. Ldr. Lawrance and Col. Cross took the salute. Following its dedication by Lt. Col. Appleby, the Aldershot Area Chaplain, the Army flag was raised. From that day, RAF Abingdon became Dalton Barracks, named after a Quartermaster who was one of eleven awarded the VC at Rorke's Drift during the Zulu Wars in 1879.

During 1993 permission was granted for the RAF to build accommodation and hangars for a Volunteer Gliding School on the airfield, away from the main site occupied by the Army. The buildings were designed in 1994 and in 1995 a new hangar for five Vigilant motor gliders, with associated accommodation, was built at the Shippon end of the airfield. In June, the RAF once again had a presence on the airfield when 612(V)GS moved in from Halton, Benson being its parent unit.

After some fifty years of gravel extraction, in 1996 the former site of Stanton Harcourt airfield featured in the Channel 4 'Time Team' archaeological exploration series. Aeronautical artefacts were not the goal but the search of a gravel pit for clues to the nature of the area's landscape 250,000 years ago! Although the runways disappeared many years ago, the hangars and many of the buildings remain on the airfield and in the village.

OFFICIAL AND LOCAL NAME - ABINGDON

COUNTY:	Oxon (since 1974 - was Berskhire)
LOCATION:	just Northwest of town of Abingdon.
LANDMARKS:	City of Oxford: 6 mls. NE
O/S-GRID REF:	SU475992 (centre of runway)
LAT:	51° 41' 30" N
LONG:	01° 19' 0" W
CONTROL TOWER:	(a)"Villa'-type to drg. 5845/39 (b) to drg. 2548c/55
HEIGHT ASL:	245 ft
LIGHTING:	Mk.II

AIRFIELD CODE:	AB
OBSTACLES:	Boar's Hill (535 ft.) 1.5 mls. NE
LANDING:	01/19 2000ft. x 150ft. concrete
	09/27 4800ft. x 150ft. concrete
HOUSING:	Permanent
HANGARS:	1 type C; 4 type A; 1 F type in 1959
OPENED :	1932
CLOSED:	1992 (except for powered gliders)
CURRENT USE:	Army

OFFICIAL AND LOCAL NAME - STANTON HARCOURT

COUNTY:	Oxfordshire
LOCATION:	6.5 mls. west of Oxford
LANDMARKS:	City of Oxford
O/S-GRID REF:	SP410050
LAT:	51° 41' 30"N
LONG:	01° 24' 30"W
CONTROL TOWER:	To drg. 343/43
HEIGHT ASL:	230 ft
LIGHTING:	Mk.II
AIRFIELD CODE:	ST

OBSTACLES:	Wytham Hill (539 ft.) 3 mls. NE
LANDING:	06/24 4800ft. x 150ft. tarmac
	12/30 3300ft. x 150ft. tarmac
	00/18 3300ft. x 150ft. tarmac
HOUSING:	None
HANGARS:	1 type T2; 1 type B1
OPENED :	9.40
CLOSED:	11.45
CURRENT USE:	Gravel pits.

UNITS PRESENT AT ABINGDON (and satellite)

UNIT	CODE	FROM	DATE IN	DATE OUT	TO	AIRCRAFT USED
40 Sqn.	OX	Upper Heyford	8.10.32	2.9.39	Betheniville(France)	Gordon; Hart; Hind; Battle
Stn. Flt. [note 1]	T5	(formed)	27.10.32	?	(disbanded)	Atlas; Moth; 504N; Tutor; Magister; Hind Whitney Straight; Don; Mentor; Proctor; Tiger Moth; Lysander; Battle; Aldon; Meteor
Oxford UAS		Upper Heyford	3.11.32	9.39	(disbanded)	Atlas; 504N; Tutor; Tiger Moth
15 Sqn.		Martlesham Heath	1.6.34	2.9.39	Betheniville(France)	Hart; Hind; Battle
104 Sqn.		ex 'C' Flt. 40 Sqn.	7.1.36	21.8.36	Hucknall	Hind
98 Sqn.		(formed)	17.2.36	21.8.36	Hucknall	Hind
52 Sqn.		(formed)	18.1.37	1.3.37	Upwood	Hind
62 Sqn.		(formed)	3.5.37	12.7.37	Cranfield	Hind
802 Sqn.		HMS Glorious	4.11.37	17.1.38	HMS Glorious	Nimrod; Osprey
825 Sqn.		HMS Glorious	4.11.37	2.2.38	HMS Glorious	Swordfish
185 Sqn.		(formed)	1.3.38	1.9.38	Thornaby	Hind; Battle
106 Sqn.		(formed)	1.6.38	1.9.38	Thornaby	Hind; Battle
103 Sqn.		Usworth	2.9.38	1.4.39	Benson	Battle
HQ Flt., AASF		(formed)	24.8.39	12.9.39	(rvtd to 1 Group)	Magister
Comm. Sqn., RAF Component, BEF		(formed)	25.8.39	16.9.39	Laval (France)	Tiger Moth; Hart; Magister, Cierva C.40
6 Gp. Comm. Flt.		?	5.9.39	11.5.42	(became 91 Gp. Comm. Flt.)	Oxford; Hornet Moth; Puss Moth; Anson; Wicko; Eagle II; Hawk; Tiger Moth
63 Sqn.		Upwood	7.9.39	17.9.39	Benson	Battle; Anson
4 Gp. Pool [note 2]		Benson	17.9.39	8.4.40	(became 10 OTU)	Whitley; Anson
1 Gp. Pool [note 3]		Cranfield & Benson	18.9.39	8.4.40	(became 12 OTU at Benson)	Anson; Battle
1 CRU (Cowley)		(formed at Cowley)	9.39	.40	Cowley	(used Abingdon as flying base until Cowley opened)
6 Gp. Target Towing Flt.		(formed within Station Flt.)	27.12.39	6.3.40	(duty taken over by	Lysander; Battle SF Bicester)
6 Gp. Comm. Flt.		(formed)	by 4.40	1.1.43	Topcliffe	(various)
10 OTU	JL, RK, UY, ZG	(formed ex 97 and 166 Sqns.)	8.4.40	10.9.46	(disbanded)	Whitley; Anson; Martinet; Wellington; Hurricane; Lysander; Defiant; Spitfire; Hornet Moth; Leopard Moth; Moth Minor; Magister; Tiger Moth; Taylorcraft
150 Sqn.	DG	Houssay (France)	15.6.40	19.6.40	Stradishall	Battle
103 Sqn.	PM	Souge (France)	15.6.40	16.640	Honington	Battle
7 AACU det.		Castle Bromwich	6.40	?	Castle Bromwich	(various)
1 BAT Flt.		(formed)	12.1.41	10.41	(became 1501 BAT F)	Whitley; Anson
1341 (SD) Flt.		(formed)	1.6.41	.44	Acaster Malbis	Whitley
1501 BAT Flt. [note 4]		(ex 1 BAT Flt.)	10.41	15.11.43	(disbanded)	Whitley; Oxford
43 Gp. Comm. Flt.	IB	(formed)	by 1941	1.1.46	Henlow	(various)
91 Gp. Comm. Flt.	M7	(formed ex 6 Gp. CF)	11.5.42	9.4.47	Swinderby	Anson; Tiger Moth; Monarch; Proctor; Cub
1682 B(D)T Flt. [note 5]	UH	(formed)	1.7.43	26.2.44	Enstone	Tomahawk
6 AACU det.		Castle Bromwich	8.3.43	?	Castle Bromwich	(various)
1341 SD Flt. (note 5)		(formed)	1.6.44	.44	Acaster Malbis	Whitley

525 Sqn.	WF	Membury	31.10.46	1.12.46	(disbanded)	Dakota
130 Gliding Sch.		Cowley	9.11.46	20.4.51	Weston-on-the-Green	(various gliders)
238 Sqn.	FM	(re-formed)	1.12.46	5.11.48	(disbanded)	Dakota; Horsa; Oxford
46 Sqn.	XK	Manston	16.12.46	24.11.47	Oakington	Dakota; Horsa
Oxford UAS	RUO	(reformed)	?	14.4.49	Kidlington	Harvard
4 Gp. Comm. Flt.	M8	Rufforth	26.4.47	2.2.48	(merged into 38 Gp. Comm. Flt.)	Proctor; Oxford; Dominie; Anson
27 Sqn.		(re-formed)	1.11.47	24.11.47	Oakington	Dakota
30 Sqn.	JN	(re-formed)	1.11.47	24.11.47	Oakington	Dakota
40 Sqn.		(re-formed)	1.12.47	15.6.49	Bassingbourn	York
59 Sqn.	BY	(re-formed)	1.12.47	25.6.49	Bassingbourn	York
242 Sqn.	KY	Oakington	1.12.47	15.6.49	Lyneham	York
51 Sqn.	MH	Waterbeach	8.12.47	25.6.48	Bassingbourn	York
47 Gp. Comm. Flt.		Little Staughton	15.4.48	1.11.49	(redes. 46 Gp. CF)	(various)
TC Dev. Unit		Brize Norton	30.6.49	14.10.51	(became TC Dev. Flt.)	Anson; Devon; Halifax Hastings; Valetta; Hadrian; Hoverfly
Air Transport Trg. & Dev. Centre		Brize Norton	2.7.49	?	? Old Sarum	?
46 Gp. Comm. Flt.		(ex 47 Gp. CF)	1.11.49	31.3.50	(disbanded)	Dakota; Anson; Proctor
27 Sqn.		Netheravon	1.3.50	10.6.50	Netheravon	Dakota
Air Tpt. Dev. Flt.		(formed)	1.3.50	?	?	?
1 Para. School		(ex 1 P&GTS at Upper Heyford)	10.6.50	1.11.53	(redes. 1 Para. Trg. School)	Dakota; Horsa; Oxford
30 Sqn.	JN	Oakington	27.11.50	2.5.52	Benson	Dakota; Valetta
Overseas Ferry Unit	QO	Chivenor	19.3.51	1.2.53	(disbanded)	(various)
TC Trg. & Dev. Flt.		(ex TCDev. Unit)	14.10.51	16.1.56	Benson	York; Hastings; Valetta; Anson; Devon
Ferry Trg. Unit		(reformed)	5.8.52	9.4.53	Benson	(various)
3 (LR) Ferry Unit		(formed)	3.11.52	1.2.53	(became 167 Sqn.)	(various)
TC Air Supp. Flt.		?	1.53	14.9.54	(became 1312 Flt.)	Hastings; Valetta
147 Sqn.		(ex 1 LRFU)	1.2.53	4.5.53	Benson	
167 Sqn.	QO	(ex 3 LRFU)	1.2.53	16.4.53	Benson	Harvard; Meteor; Valetta
24 Sqn.		Topcliffe	7.5.53	1.1.57	Colerne	Hastings
47 Sqn.		Topcliffe	13.5.53	31.12.67	(disbanded)	Hastings; Beverley
1 Para. Trg. Sch.		(ex 1 Para. School)	1.11.53	31.12.75	Brize Norton	Hastings
1312 (Transport Support) Flt.		(re-formed ex TC Air Support Flt.)	14.9.54	1.4.57	(disbanded)	Hastings; Valetta; Chipmunk
53 Sqn.		Lyneham	1.1.57	28.6.63	(disbanded)	Beverley
Air Movements Dev. Unit		(formed)	11.59	31.5.65	(disbanded)	?
Andover Trg. Flt.		(formed)	1.7.66	9.9.70	Thorney Island	Andover (borrowed)
46 Sqn.		(re-formed)	1.12.66	9.9.70	Thorney Island	Andover
Air Support Cmd. Exam. Unit		Benson	13.11.70	1.9.72	(redes. 46 Gp. Air Trans. Exam. Unit)	?
Joint Air Transport Establishment		Old Sarum	4.71	22.7.76	Brize Norton	?
46 Gp. Air Trans. Exam. Unit		(ex AS Cmd. Exam. Unit)	1.9.72	31.10.73	Upavon	?
6 AEF		White Waltham	23.8.73	31.7.92	Benson	Chipmunk
ULAS		White Waltham	10.8.73	31.7.92	Benson	Chipmunk; Bulldog
OUAS		Bicester	26.9.75	31.7.92	Benson	Chipmunk; Bulldog
612 (V) GS		Halton	7.6.95		(current)	Vigilant

Note 1: wartime code allocation; no evidence of use
Note 2: comprised 97 and 166 Squadrons from Benson
Note 3: comprised 35 and 207 Sqns. from Cranfield and 52 and 63 Sqns. from Benson. 207 Sqn. left 9.12.39 and 35 Sqn. left 1.2.40.
Note 4: used Stanton Harcourt satellite from 18.4.43 to disbandment
Note 5: used Stanton Harcourt satellite